AF394569

ROBIN PRYTHERCH
A LIFE WITH BUZZARDS

Compiled and edited
by
Lyndon Roberts

Bristol Books CIC, The Courtyard, Wraxall,
Wraxall Hill, Bristol, BS48 1NA
www.bristolbooks.org

Robin Prytherch:
A Life with Buzzards
Edited by Lyndon Roberts

Illustrations by Robin Prytherch

Published by Bristol Books 2024

ISBN: 978-1-909446-42-7

Design: Joe Burt

Printed on paper from a sustainable source.

A CIP record of this book is available from the British Library

Editor's Foreword	7
Robin Prytherch (1939–2021)	11
Introduction	15
The Buzzards	19
Bibliography	73
Buzzard territories	74
Index	76
Acknowledgements	79
A note about our sponsors	80

Dedicated to the buzzards and
landowners of North Somerset who
made Robin's studies possible.

EDITOR'S FOREWORD

I first met Robin Prytherch in the early 1990s, by which time his study of common buzzards was well under way. We shared a common interest in bird ringing (banding), and it was not long before I was persuaded to help him ring nestling buzzards. The arrangement was that he would locate the nests and climb the trees and I would be his groundsman. The reward for my efforts was a bagful of young buzzards lowered to me on the ground at the end of a rope.

Robin was very methodical and thorough in his search for buzzard nests. The search itself was no random affair; it was the culmination of many hours of meticulous fieldwork, observing buzzards' behaviour, mapping their territorial boundaries and making a note of the likely nest sites that he saw. He was particularly careful to note where adult birds made a beeline for a likely nest site (sometimes carrying nesting material or prey) then check out the locations at closer range. Some nests were relatively easy to find using a telescope or just a pair of binoculars, especially if the right vantage point was chosen – something Robin was very good at. However, finding most required more legwork and fighting through woodland understorey – always with the landowner's permission.

Once located, the nest was observed from a distance to determine which stage it was at, from nest-building through to incubation, hatching and rearing. Usually, Robin could judge if it was the right time to ring the chicks by observing their size, the amount of down on their heads, and the size and spread of the droppings beneath the nest tree. If the chicks were too small, the ring might slip off, possibly causing injury in the process. If the chicks were too big, there was a danger of they might vacate the nest prematurely. When he was satisfied the time was right, we would visit the nest site together, armed with ladders, rope and other paraphernalia to get Robin up to the nest. He was a very good climber and, although some nests proved frustratingly difficult to reach, he was usually up at the nest within 30 minutes of arrival. A few we had to give up on.

On arrival at the nest site, Robin would climb the tree and tie himself to the trunk or a sturdy branch for safety. He would then place the chicks in a small rucksack and lower them to me on the ground. I then attached a uniquely engraved/numbered metal ring to the right lower leg (the tarsus) of each chick. The rings were supplied by and registered with the British Trust for Ornithology (BTO). After ringing, several biometric measurements were taken, including wing length and mass. At this stage, the wing feathers would still be growing but were a useful indication of age (in days since hatching). After ringing and measuring, the chicks were placed back in the rucksack to be hauled back up the tree by Robin, who carefully placed them back in the nest, exactly as he had found them. Quite often, while all this was happening, he would describe what he could see from high up in the tree canopy, sometimes to my mild annoyance! In later years, he began taking detailed notes, describing the structure, composition and orientation of the nests. He also recorded the prey remains he found within them.

Normally, while we were at a breeding location, the parent birds would just quietly slip away. Some would circle high over the tree; others would settle in a nearby tree, just watching and waiting. Some (though not many) were more demonstrative, but all would quickly return to their offspring once we had left. It always fascinated Robin that parents behaved unpredictably when their nest sites were disturbed in this way, and it added to his understanding of the birds in his study area as individuals, to be named and illustrated.

There was a lot more work to do after ringing the chicks and vacating the breeding site. Robin would revisit the territory frequently, observing and listening for the signs that would tell him how many chicks had fledged and how long they had stayed in their natal area before being chased out by their parents. In some cases, he even documented how they perished – the first few months after fledging are a perilous time for young buzzards.

At the end of each breeding season, I submitted data to the BTO. Robin was not digitally minded, however, so he undertook the task of filling out BTO *Nest Record Cards* by hand in his very neat handwriting.

Robin's interest in buzzards did not only cover the breeding season; he was interested in their whole life cycle. To record the individual life histories of the birds in his study area, he spent long hours in the field, mostly working alone. He often logged more than a thousand hours each year, watching his buzzards, noting their behaviour and sketching their antics as well as their plumage features, postures and interactions with other wildlife. All these observations were recorded meticulously in his 187 field notebooks. He was well known by local landowners and farmers in his study area as "the buzzard man". He was also the recipient of much useful information about his birds, including those found dead or moribund. During the winter months especially, he was popular on the talks circuit and used that medium (as well as print, radio and television) to spread the word about what he had discovered and what questions he yet hoped to answer.

There is a part of North Somerset that will feel empty without the buzzard man and no doubt his absence will have been noted by the birds he studied there. However, the buzzards will live on, we hope, until – and after – someone else comes along to pick up where Robin left off.

Lyndon Roberts

RJP holding a hobby *(Falco subbuteo)* nestling.
Nailsea, North Somerset, 1989.

Photographed by Ken Hall.

ROBIN PRYTHERCH
(1939–2021)

Most of the content of this book illustrates just one aspect of Robin Prytherch's many and varied interests, although it is perhaps the one most of his friends would particularly associate him with: his studies of the common buzzard.

So, who exactly was Robin? The raw facts are that he was born in Sussex in 1939 and spent his childhood there, or at school in London, before working as a technical draughtsman and designer for an engineering firm in Enfield. His interest in natural history had already been awoken by then – the story goes that the inspiration for his name was from a bird singing nearby when he was born.

In the mid-1960s he moved to Bristol, which became his permanent home from then on. In 1968 he joined the BBC's Natural History Unit – initially as a researcher then gradually becoming more and more involved in film production, working closely with many of the names with whom we are familiar from our TV screens. In fact, his expertise, particularly as concerned bird identification and behaviour, was highly valued there, and he was often the 'go-to' person to provide answers when technical queries arose, or if background information was needed.

On arrival in Bristol, he soon joined the committees of the naturalists' societies for both Bristol and Somerset. However, frustration with the limited resources available to them by the then numerically dominant birdwatchers led Robin and a few others to form the Bristol Ornithological Club (BOC) towards the end of 1966. This grew rapidly, producing a monthly newsletter – as it still does – and organising indoor meetings, field trips and surveys. Robin, always a key player, was elected chairperson from 1972–1975, remaining on the club's committee until 2013. He was one of only two Honorary Life Members in the club's history. Not only that, but he also designed the BOC's logo – a neat depiction in a few lines of a pied-billed grebe. This is a North American waterbird he identified at Blagdon Lake in 1963, thereby adding a new species to the British List. It was the bird of the moment at the time of the club's formation.

In the years that followed, until his death in 2021, he led hundreds of trips for the club, including to the Netherlands and Majorca, but he was also very happy to act as a back-up for others. Robin loved to travel with friends, considering the reduced carbon footprint per person an extra bonus. Although, just like any other member of the public, he enjoyed looking at birds for no other reason than the pleasure they provide, he was also keen to promote their more serious study – 'purposeful birdwatching', as it is sometimes termed. To this end he organised various local studies, most notably of the more common birds of prey in our area: kestrel, sparrowhawk and buzzard. He mapped the results in the club's journal, *Bristol Ornithology* (see Bibliography), of which he was an editor throughout its existence.

Robin's own buzzard studies were focused on a 75 square kilometre patch of farmland west of Bristol, stretching east from Clevedon, through the Gordano Valley as far as the outskirts of Portishead, and south through Failand, as far as Barrow Gurney. Here, he could be found throughout the year, mapping the ever-increasing number of territories – rising from 13 in 1982 to an amazing 120 in 2020 – but also carefully observing the birds' social and breeding behaviours, their choice of nest sites, and the range of prey items they brought to their young. He estimated he spent up to 1,200 hours out in this patch of countryside every year, each day's visit lasting between three and eight hours on average. He was there every month, although most frequently in spring and summer when pairs were displaying, nest-building and feeding young. He maintained good relations with the locals, becoming well known to quite a few of them, including the farmers and gamekeepers on whose land the birds were nesting.

He had been a bird-ringer from early on in his birdwatching career, being a mainstay of the ringing station at Chew Valley Lake for many years. He was particularly committed to training newcomers to this aspect of scientific bird study. So, it was natural that, as part of his buzzard work, Robin ringed as many chicks as he could, to try to find out where they went and how long they lived. Watching him swinging from a flexible ladder could be quite unnerving and he remained an agile tree-climber well into middle age. In due course, his results appeared in three key papers in the journal *British Birds* (see Bibliography), illustrated by his own line drawings, with other shorter papers and comments published elsewhere.

As far as buzzard identification was concerned, Robin had an uncanny ability not only to distinguish males from females and adults from juveniles, but was also able to recognise individual birds, primarily by their plumage, and in some cases by their voice, 'personality' and shape. This made following the life histories of each one that much easier. He then used his skill in pen-and-ink line drawing to sketch each one as an aide-mémoire. These are the images, along with the accompanying texts, that were incorporated into the much-loved Christmas cards he sent to his friends each year, and which form the heart of this volume.

Ken Hall

A more detailed account of Robin Prytherch's life, including his many achievements and interests, can be found in the journal 'British Birds' (Vol. 114, issue 5). An obituary was also published in The Guardian on 15.03.21 – Ed.

NOMENCLATURE

The term 'buzzard' here refers to the nominate (British and western European) subspecies of the common buzzard (*Buteo buteo buteo*). It should not be confused with the turkey vulture, which is often referred to as a buzzard in North America. The common buzzard's closest relative in North America is *Buteo jamaicensis* which, to confuse matters further, is named the Red-tailed Hawk.

The standard convention in scientific papers, including those authored by Robin Prytherch, would be to use initial capital letters when referring to the Common Buzzard. However, to help with the flow of the narrative, the choice has been made here to use the lower case version, 'common buzzard' for this and other species mentioned in the text.

INTRODUCTION

After Robin Prytherch (RJP) died in 2021, there was much celebration of what he had accomplished among those who knew him in life and those who had read and heard about his achievements – about which he was very modest. What he has left behind includes a wealth of written material about birds and his wildlife illustrations. As a friend, however, I felt that something was missing.

Robin's real passion as an ornithologist was for buzzards and he devoted much of his life to studying them. A trait I quickly learned to recognise in him was his remarkable ability to observe (and interpret) behaviours in the field that eluded the casual observer, or even more experienced birders. He delighted in finding new and interesting things about buzzards and, I believe, it was something he loved to share with others. So, starting in 1995, he decided to do just that by designing and sending out his own buzzard-themed Christmas cards. In my view, those cards (collated here) represent the missing piece in RJP's legacy. Not only do they complement his serious, scientific publications; they also stand alone as a more personalised account of his endeavours to understand the lives of buzzards.

The Christmas cards make up the bulk of this small book. There were 26 in total, covering the period 1995 to 2020. Each contained an illustration by RJP on the front (usually print-copied from a pen-and-ink drawing, although some were pencil sketches), with accompanying text on the reverse. The central pages were blank, except for Robin's handwritten personal greeting. Robin was always very particular about them; they had to be perfect. As he was a stickler for consistency, the size was always the same: A6 (folded A5), although the orientation varied – sometimes portrait and sometimes landscape. They were always printed against a plain, pale-coloured background, often beige but sometimes a subtle shade of blue, green, mauve or grey. Nothing too showy.

Although not exactly festive, the cards were a delight to receive. The front covers displayed a real talent for detailed illustration and the comments on the back were insightful. They provided a window on his world, a world he wanted to reveal to his network of friends, colleagues and fellow ornithologists. Of particular interest was the way in which he named individual buzzards and followed their progress in life and, in some cases, their demise. Each individually named buzzard was recognisable to him because of expertly observed plumage variations. The cards represent a very personalised view of his subject matter, not just because they were based on his own observations and drawings, but because they dealt with buzzards as individuals to be recognised and studied, almost as if they were humans.

Collectively, the cards cover a broad spectrum of events in the life cycle of a buzzard, from pair formation, nesting and the raising of young to old age, infirmity and death. They also describe a range of more salacious behaviours, including bigamy, incest, fratricide, piracy, evictions, takeovers and divorce. As RJP commented himself, whoever thought the lives of buzzards could be so complicated? Great gatherings of buzzards are illustrated and described. 'Angel' postures, display stoops and aerial play are explained, as well as less-known behaviours such as cooperative breeding.

Robin Prytherch was, without doubt, an expert in his field. His writings on buzzard behaviour and breeding ecology stand out as landmark papers in these fields of study. He was fortunate to be able to study buzzards in one relatively small area of North

Somerset, a short distance from his home in Bristol (England). Through his studies, he demonstrated that North Somerset holds what is believed to be the highest-recorded density of buzzards within their known breeding range. It is a remarkable story, and hopefully this book will inspire others to undertake in-depth studies of wild birds so that we might understand them better.

The cards reproduced here are an important part of his legacy.

Lyndon Roberts

THE
BUZZARDS

OLD GIRL

I first saw this female buzzard on 7th September 1980, near Flax Bourton, when she was moulting out of her juvenile plumage. She was about 16 months old. A year later, she joined a male who had lost his mate that spring. This territory, which I called 'Belmont House', was adjacent to the area where I'd first seen her. Since then, I have followed her life history and only failed to record her breeding success in two years. In her life, Old Girl fledged at least 16 chicks with four mates. Her first mate may have been quite old. He was a beautiful, almost-black buzzard named Black Beauty. In 1987, he was replaced by New Boy, who survived for six years. He was then replaced by Hal, who survived for only one year. Her next mate was in the territory for only a few months. The last time I saw her was on 16th September 1995 with yet another male, Vig. In October, Vig was with another female, so Old Girl had almost certainly died in late September. She had lived for 16 years and about four months – probably a good age for a wild buzzard.

December 1995

BUFF

This rather pale female buzzard lived in Hack's Wood territory, which included most of the wood and the meadows below. The sketch shows her hunting for worms in a field of recently seeded grass. I first identified her in 1986, but a pale bird – probably her – had been in the territory since at least 1983. When she died in January 1996, she would have been at least 12 years old, but probably at least 15 and a half. During the 10 years from 1986 to her death, she was mated to Rufous and they fledged 16 chicks (but they probably fledged others in the three previous years). 1993 was the only year in which I know Buff and Rufous failed to fledge a chick. That autumn, they were joined by another female. This bird was about 15 months old – in its second year, and in its first adult plumage. It is very rare for a pair to accept a stranger into their territory, but this one was certainly tolerated. Was this to be my first record of a bigamous trio? During the next two years, this bird stayed in the territory but did not have its own nest. Instead, she showed some interest in Buff's nest and subsequently helped to provision the chicks with food. After Buff died, this female, now named Nan, took her place, and this year Rufous and Nan fledged one chick.

December 1996

BUZZARDS AT PLAY

During their first year, most young buzzards roam widely, travelling around as singletons, or in twos or threes. When not concentrating on the search for food, they will often consort in aerial play. This will mostly take the form of dives and chases; the action is never in earnest but is intensified if one of them carries a small object in its talons (mock prey), as shown top left. Obtaining such items forms another part of play; birds will swoop down into trees to grab at twigs or cones. Quite frequently, the object does not break off and the young buzzard finds itself upside-down, wings flapping wildly to right itself (top right). As soon as the object is secured, it is carried aloft, only to attract the attention of other birds nearby. The object may then be dropped, but – frequently – the buzzard will dive down to retrieve it (lower right), unless another one swoops down and steals it. When on the ground, perhaps when hunting for worms, the sight of a small clod of grassy soil will stimulate the buzzard to pounce on it at lightning speed (lower left). The clod may then be held as the bird jumps with it, releases it, pounces again and shakes it. These activities are all part of a young buzzard's training for adult life, when chasing intruders from its territory will be in earnest and when hunting efficiency could well determine its breeding success, if not survival.

December 1997

CURT

Curt and his mate Cora settled into their territory in early 1991. Situated between the M5 and A369, just east of Clapton in Gordano, it is a territory without woodland (apart from a small copse occupied by rooks). However, there are plenty of hedgerow trees and this is where they have nested, usually in willows. They failed to breed successfully in their first year, then fledged five chicks over the next three years. In March of the next year (1995), Curt damaged his left eye. For most of the time it was shut, but he usually opened it as he took flight. I wondered, how long he could last with this crucial part of his hunting equipment out of action? Breeding attempts failed that year and for the next two, during which Curt's eye appeared to be out of use. By last winter it was 'very shut' and even appeared sunken – he was now definitely blind in one eye. Nevertheless, he seems to have mastered his disability because, throughout the period, he was active in the defence of his territory. His hunting ability was, no doubt, affected to some degree and could explain the breeding failures, but he is obviously back in form; this year, he and Cora fledged one chick. This astonishing and possibly unique event in the life of a wild buzzard, heavily dependent on sight for survival, is testament to this bird's overall fitness.

December 1998

Robin Prytherch 1998

HAZE

The sudden break-up of Haze's territory early this winter alerted me to the possibility that either Haze or his mate, Ivy, had died. Three new buzzards – a pair and a lone bird – had settled into parts of the territory. Normally, Haze or Ivy would have chased them out, as buzzard pairs are very territorial. However, despite hours of watching, I have not seen them since the new birds arrived. It is a good territory, which includes substantial woodland, grass paddocks and hedgerows. Haze first appeared there in 1989 and was joined by a very pale female. During the next spring, a new dark female, Ivy, took over and the pair remained together until now. Haze, with his two mates, has fledged 17 chicks in his life. This is an above average breeding performance (just over 1.5 chicks per year) when compared with all the breeding birds in my study area over the same period (an average of just over 1.1 chicks per year). If he arrived in the territory as a young bird at three or four years old – which is likely – then he has lived for 14 or 15 years, breeding successfully in eight of them. He was named Haze because his plumage always appeared washed out. The sketch shows him hunting from a favourite perch in a large dead tree. This also served as a look-out to watch over the territory. The same tree now appears to be being used by the new pair.

December 1999

Robin Prytherch, 1999.

SOOTY AND SNOWY

The exceptional variation in the plumages of buzzards has enabled me to follow individual birds throughout their breeding lives, and sometimes from their first year. The two extremes of dark and pale birds are demonstrated well in my study area by a dark male, Sooty, and his pale mate, Snowy, as shown in the sketch. Another pair, Nero and Mask, are similar, but Mask is slightly darker, with a creamy wash to the white areas, whereas Snowy is purer white. In other pairs, the male is normally the paler of the two, or there may not be much difference. Between the two extremes there is infinite variation, and even in a pair of very similar birds there will be a small difference, say, a pale chin in one and a dark chin in the other. The mostly brown colours vary from dark purplish, or blackish brown to pale buff, cream or white, with occasional rufous or grey (particularly on the tail).

Sooty arrived in his territory with his first mate, Larch, in 1993. By 1999, they had fledged just six chicks – a poor breeding performance. Her plumage was average, but distinctive. Then, in early spring this year she disappeared and Snowy arrived to take her place. Sooty and Snowy attempted to breed this year, but failed. She is likely to be only in her fourth or fifth year and even though Sooty is probably 10 or 11 years old, they could have many successful years ahead.

November 2000

Robin Prytherch, 2000

HACK'S WOOD

This card shows two generations of buzzards in Hack's Wood territory. Rufous (top left) with his second mate Nan (top right), and below, the new pair (from October 2000) Hak (male, left) and his mate Cal. The quiet scene shown is imaginary; the two pairs probably never met at close range.

Rufous's first mate, Buff, has featured on an earlier card. With her, Rufous fledged at least 16 chicks between 1986 and 1995, and then four more with Nan until they both disappeared in the autumn of 2000, presumed dead. Rufous was already well established in the territory with Buff in 1986 and had probably been there for at least the previous three years. When he died, he was probably around 20 years old – a long-lived buzzard. The 20 or more chicks he fledged in his lifetime represent moderately good productivity. Nan's contribution, however, was poor – just four chicks in her eight years, although she did help Rufous and Buff by providing food for their last two broods. I think it is highly likely that Nan was one of Rufous and Buff's chicks (the one fledged in 1992). This means that Rufous was mated to his daughter – an unusual event, but one which is known to occur in birds. This may explain the low productivity. Rufous was, of course, getting old and this could be another reason. Hak and Cal fledged their first chick from the same nest Rufous and Nan used in the previous year.

November 2001

Robin Prytherch, 2001.

CHOCOLATE AND CREAM

Westpark Wood buzzard territory, which straddles the M5 on the south side of the Gordano Valley, has been occupied throughout my study period (since 1982). The original, unnamed pair were together until the female died in early 1991. She was replaced by a rather pale bird, Cream (left). The original mate died on or soon after 11th January 1993, on which day I saw him looking rather groggy. He allowed me to approach within a metre, but eventually flew off rather weakly. Just five days later, a new male was displaying over the territory. His plumage was predominantly dark brown, and Chocolate seemed an appropriate name (right). During their nine years together, Chocolate and Cream fledged a total of 14 chicks (although Cream had already fledged four chicks with her first mate). Early in the breeding season this year, Chocolate must have died, as a new male appeared. I first saw him with Cream in mid-May, when it became clear that this year's breeding attempt had failed. Chocolate had lived for 12 or 13 years, which is far from old age for a buzzard. Cream lives on with her third mate, Cuss, also a rather pale bird, and even though she is now 14 or 15 years old, she could survive for five or more years yet.

November, 2002

Robin Prytherch, 2002.

TESS

Tess has been the most successful breeding buzzard in my study area, fledging an astonishing 43 chicks over her life. She first appeared in her territory at Clapton Wick in the Gordano Valley in 1984. Tess paired up with a male, Tertials, who established the territory the previous year. Their first breeding attempt (in 1984) failed – a single chick hatched but died soon after. In every year thereafter, except 1989 and 2003, she bred successfully, fledging 13 of the chicks with Tertials. He died in the spring of 1991 after he engaged in several clashes with another male. The remains of Tertials were found in the territory. By then, the other male, which I named Tuff, was already with Tess. He and Tess went on to fledge the other 30 chicks – Tuff is obviously an excellent male. The 43 chicks were from just 17 broods (three broods of four chicks, six of three, five of two and three of one). Only one other pair in my study area has fledged a brood of four, to my knowledge. Tess was last seen this spring when she would have been in her 22nd or 23rd year (assuming she was in her third or fourth year in 1984). This autumn, a new female was present with Tuff, who must now be about 16 or 17 years old. The sketch shows Tess with one of her long-staying juveniles, calling just beyond, and Tuff on a favourite look-out perch in a poplar in the distance.

November, 2003

Robin Prytherch, 2003

FLAX AND FLOSS

This pair of buzzards has a territory close to the village of Flax Bourton. The male, Flax, is in the centre, with his mate, Floss, near left. Unusually, Floss was in her first year when she paired up with Flax, who was at least three years old. Buzzards don't normally pair until they are three years old or more. Although Flax built nests in the first two springs, breeding activity did not continue, and it was only when Floss was three that she bred. However, the nest failed, as it has in all but one of the four years since. In the successful year, 2002, the pair fledged just one chick. This is a poor result from five breeding attempts. Interestingly, the chick, now in its third year, is still with its parents; in the sketch it is just beyond Floss.

This autumn, there was a great gathering of mostly non-breeding buzzards in two large autumn-sown fields near Flax Bourton, partly within the pair's territory. They had settled to feed on the abundant earthworms. At the peak of the gathering,* at least 60 buzzards were present (a few are shown in the sketch). The pressure of numbers meant the pair could not evict the intruders, although Flax and Floss frequently attacked them. Here, Floss adopts a dominant posture, having just displaced the bird to her right.

December, 2004

---

*The collective noun for buzzards in the United Kingdom is a 'wake' – Ed.

Robin Prytherch. 2004

INDI AND ABBY

Caswell Cross, the territory where this pair live, has had an unusual history. It was first occupied in 2000 by another pair, Cas and Tear. In the next territory to the west, the male, Como, lost his mate, Cora, late in 2001. Then, in March 2002, Tear started to visit Como regularly; indeed, she mated with both males, and eventually settled with Como – my first evidence of a buzzard divorce! Cas then attracted a young female, Abby, who had been living alone in a small territory to the east. However, within about two weeks, Cas had been evicted (or killed) by a new male, Indi. The new pair, Indi and Abby, are still in the territory, but so far have fledged only one chick (in 2004) – a rather poor breeding performance. I first saw Abby when she was in her second year, so I can be sure she is now in her eighth year. Indi is probably of a similar age.

In the sketch, Abby is eating a black-headed gull. Indi waits and watches; females usually dominate males at prey. Magpies and carrion crows will try to steal titbits – in this case a magpie has succeeded. Although small mammals, including voles, mice and rabbits, are an important part of the buzzard diet, so are birds – especially corvids and pigeons, but also thrushes and even tits. Earthworms will also be readily taken. Some prey is stolen from other raptors (I've seen thefts from sparrowhawk, kestrel and peregrine) or picked up as carrion.

November 2005

Robin Prytherch, 2005

MASK AND NERO

This pair of buzzards settled into their territory, known as Court Hill, in 1997. It covers the woodlands above Clevedon Court and south over the M5 to Clevedon Moor. In the first year, they fledged a single chick, but failed for the next four years. Then, in the following five years, they fledged 10 chicks – a moderate breeding performance over 10 years. Later this year, Mask disappeared and is presumed dead (at just 12 years old – middle-aged for a buzzard), as Nero now has a new young mate, in her third year. This was the age at which Mask first bred, and I know this because I first saw her when she was in her first year, eight kilometres to the east near Flax Bourton, on 16th October 1994. She was amongst a group of 24 non-breeders (almost all were her age) on stubble and/or freshly ploughed fields, where they were feeding on worms and other invertebrates. She had a strikingly dark face, and in my notebooks, I described her as 'smudgy face'. The resident pair there had a constant battle to keep the young birds out of their territory, and on a couple of occasions I saw 'smudgy face' chased by the female, Old Girl (shown on my 1995 card). The young Mask stayed in that area on and off for more than four months – I last saw her on 21st February 1995. It was not until 1997 that I found her again, now with her mate Nero. Where she lived in the meantime, or where she was fledged, is unknown to me.

November 2006

A BUZZARD QUARTET

Buzzards normally form monogamous pairs, but there are exceptions; the four birds featured here is one. They are, from the left, Abby, Pip, Indi and Cala – three females and one male (Indi). Abby and Indi were shown on my 2005 card. Since then, there have been extraordinary developments involving their territory, Caswell Cross, and an adjacent one, Portbury Moor. A trio of buzzards lived in Portbury Moor: the male Hun, with his mate Pip and their daughter Cala (who had remained in the territory since she fledged in 2001). During the winter of 2005/06, Hun disappeared, presumed dead. This is when I first noticed Indi flying over both Caswell Cross and Portbury Moor territories. For a long while, I was rather confused about this development. Then, as I watched Pip and Cala on the nest they shared, the problem was solved. After a change-over, Pip flew to the top of the tree, whereupon Indi dived out of the sky to settle on her and copulate. So, Indi had taken over the territory. But what about Abby, Indi's original mate? Well, Indi had not totally deserted her; she was using her own nest. The amazing fact is that he was defending both territories and the three females.

However, during 2006 and 2007, Abby did not fledge a chick, whereas Pip and Cala fledged five in the two years. With two females tending one nest, this is perhaps not surprising, regardless of Indi's input. Interestingly, when the fledged chicks attempted to stray into Abby's territory, no doubt having watched their father go that way, Abby quickly evicted them. It will be fascinating to see how this situation develops in the coming years.

November 2007

Robin Prytherch. 2007.

CHESTNUT

I've known this male buzzard for 15 years, since 1994, when he first occupied a territory at Norton's Wood, near Clevedon. His name comes from the chestnut colour on his breast and tail, with a hint of the colour elsewhere on his body. Over the first three years, he and his first mate, Willow, fledged three chicks. At the end of her third year, however, she and one of the juveniles were electrocuted at the same low-voltage transmission pole. Before the next spring (1997), a new female, Fran, had arrived and the pair are still together, having fledged 13 chicks in 12 years.

In the sketch, Chestnut (now probably in his 19th year) is just about to swallow a small earthworm. These are eagerly sought out during autumn and winter. Earthworms are easier to find in paddocks with a short sward – often those grazed by horses and particularly where moles are active, as here. Notice that his right leg seems out of place. He had obviously injured the leg somehow – I first noticed it on 5th February 2003. He walks with a limp but seems to settle on posts and branches without trouble; he also catches large prey, which is essential for large chicks in the breeding season. So, this disability does not appear to have affected him significantly. This year Chestnut and Fran failed to breed successfully, but a female juvenile (one of two chicks fledged last year) is still in the territory, now in its second-year plumage. Will it stay on and become my fifth case of a territory with a trio of adults? I shall be keeping an eye on it.

November 2008

Robin Prytherch, 2008.

PIP

Buzzards are very territorial. Pairs keep to their territories and expel any intruders. They use special behaviours to signal that they are in a territory. This will encourage intruders to move on. Here, I show a female, Pip, dealing with a juvenile intruder (based on a field sketch dated 28th September 2000). The juvenile had settled on a bush at the edge of the territory and was soon displaced by the resident pair. Then, Pip dropped down to the juvenile, which had settled in the field nearby. There was a brief scuffle when Pip displaced the juvenile before she adopted the assertive bow, head sleeked and pointing down, body horizontal with the wings flared slightly, indicating her dominance. She then sidled towards the juvenile (side-on, head towards it), resulting in it adopting an angel posture, calling very quietly – a partly submissive behaviour. Pip soon relaxed, walked away and flew back to the bush. The juvenile had stood its ground but learnt that a pair had claimed some of the field, which it should not stray into.

Pip and her mate, Hun, had settled into their territory late in 1999, close to Portbury and the M5. During the winter of 2005/06 Hun disappeared and Indi, the male in an adjacent territory, took over (as shown on my 2007 card). With her two mates, Pip has fledged 21 chicks in 10 years, an above-average breeding performance. However, unusually, one of Pip's first chicks, Cala, a female, stayed with her until 2008 and helped with raising the chicks, some of which may have been hers. Pip is now 13 or 14 years old and could have several more years to fledge yet more chicks.

November 2009

Robin Prytherch, 2009

BUFF 2

I named this female buzzard Buff 2 because she was so like another older female named Buff. I even thought that Buff 2 might be Buff's daughter – but that's all conjecture. Buff 2 settled into a territory in 1990 based on Weston Big Wood, just 2.5 km from Buff's territory in the Gordano Valley. Her mate throughout was Blotch, so-called because of uneven plumage in his wings. Although they failed to breed successfully in six of their 20 years in the territory, during the other 14 years they fledged 29 chicks, averaging 1.45 chicks per year – an above average result. They compensated for all the failures by fledging four broods of three chicks, seven of two and just three of one.

Buzzards are highly aggressive, and this behaviour starts early. In the nest, when the chicks are 8–10 days old, the older chick (they don't all hatch at the same time) may attack a younger one, sometimes killing it. This is known as 'Cainism'. I was looking through my telescope at the nest on 3rd June 2000 and I could see Buff 2 feeding two small chicks. She stopped, then moved to the side of the nest and soon one of the chicks started to chase the other around the nest, pecking at it. 'Abel' fled to the edge of the nest and 'Cain' settled in the centre. Buff 2 just watched or ignored the chicks. Bouts of chasing and attacking continued for 20 minutes until Blotch arrived with food. Rather than feed them, Buff 2 settled to brood them. The fighting only lasted a couple of days and these two chicks fledged successfully.

The sketch shows Buff 2 looking on as Blotch flies up to chase an intruder out of their territory. Sadly, the pair disappeared at the start of this year. They would have been 23 or 24 years old. That's a good age for a buzzard.

November, 2010

Robin Prytherch, November 2010

RIV AND HONEY

In 1994, a large buzzard territory at the west end of the Gordano Valley split up into four, although the smallest held just one adult male, Riv. This very small territory was only 22 hectares (a quarter of the mean of all territories in 1995). A year later, a female settled in, and they fledged one chick. The nest was in a birch festooned with honeysuckle; indeed, the nest was partly supported by the creeper – quite the sweetest smelling nest I'd visited. Honey had to be her name.

Over 16 years they fledged 21 chicks and only failed in one year – an excellent record even though brood sizes must have been limited by the small territory. Nine broods were of one chick, six of two. The sketch shows Riv feeding on a road-killed rabbit I had placed in the territory. The crows found it first, which led Riv to it. Honey soon came running in to claim the food.

Over the winter of 2010/11, Riv disappeared (no doubt he died) when he would have been at least 19 years old. In the spring of 2011, a new male appeared, but he was familiar to me as the male from an adjacent territory formed in 2010. Had he lost his mate and transferred to Honey? After hours of watching, I realised that his original mate was still in her territory; he had not deserted her, he had paired with both females – a clear case of bigamy. Interestingly, the females kept to their territories whereas the male (named Gos) flew over and hunted in both territories. They bred successfully, with Honey fledging one chick and Gos's original female fledging two.

November 2011

CRESCENT

I first saw Crescent, a female, late in 1993 in part of a large territory that was splitting up. In the next year, a mate, named Q, had arrived, and they were established in the territory at Conygar Quarry (near Clevedon). Eleven years later (in 2005), Q was replaced by a new male. Then, during the winter of 2011/12, Crescent must have died. She had lived for at least 21 years (allowing three as a juvenile/non-breeding bird before settling into the territory). Q would have been at least 14 years old.

Each year the pair had a nest, and all but one of these were sited in just two poplars in a small plantation in one corner of the territory. The first nest was used continuously for 11 years, until Q died; the second continuously for the remaining seven years with the new male until Crescent died. Both nests were visible from nearby roads and easy to check. In all of the 18 years, she failed to fledge a single chick. She seemed to have eggs, as I could see her sitting tightly on the nest. Strangely, each year, at about the time of hatching, she deserted the nest. I could only assume she was laying infertile eggs, although there might have been some other problem.

This year, I was checking as usual, but the female's behaviour was different – she was much more wary of me. Was this bird Crescent? No, it was a new female, although she looked a little like Crescent. She went on to fledge two chicks with the same male (from the same nest used by Crescent last year), despite the regular deluges of rain.

November 2012

Robin Prytherch, Nov. 2012

GOS

This male buzzard was mentioned in my 2011 card about Riv and Honey (more below). I first saw him when he joined a female in a new territory in 2009. He was a striking bird with a well barred belly and flanks, quite unlike the usual buzzard markings. The barring reminded me of a goshawk, hence the name. In their first year together, he and his mate fledged three chicks. The nest was a massive pile of twigs at the top of a small larch stunted by the closing of the canopy of other larger larches over it. In 2010, the nest failed. Then, in early 2011, Gos started to visit the adjacent territory from where Riv had disappeared. Gos's mate often watched from a high tree as he ventured into the other territory. On his return, Gos would engage in display stoops with her.

I eventually realised that he had also paired with Honey. He had not deserted his regular mate, so I witnessed the formation of a case of bigamy. Both females fledged chicks, two with his first mate and one with Honey. The bigamous relationship continued into 2012, with the first female fledging two, but Honey failed. This hectic lifestyle probably got the better of Gos as he disappeared during the winter of 2012/13 when new males arrived in both territories. Gos would have been at least seven years old (a young age for an established breeder), but he fathered eight young – a good result from just four seasons, even if from six nests.

December, 2013

Robin Prytherch Dec 2013

REV, SPECKLES AND SPOTTY

Rev and Speckles settled in a territory in early 1992. After 22 years they have disappeared, as I last saw them earlier this year. They had lived for at least 24 years – a good life for a buzzard. In the first year, their territory was based on the Avon Wildlife Trust (AWT) reserve in the Gordano Valley. Over the next two years, the pair shifted their territory between two other pairs, but it still included part of the AWT reserve. This was a time of maximum growth in the local buzzard population. They failed in their first breeding attempt, but in the next year fledged two chicks, and over the following 18 years fledged a further 18–20 in all, failing in just eight years. This productivity is about average for pairs in my study area.

In 2000 and 2002 I had views of a third adult in the territory – probably a female and very similar to Rev. Then, at the end of 2006, I saw what could have been the same bird sitting next to Rev; Speckles was sitting nearby. It was a female and probably the pair's daughter from an early brood. They became a trio; a male and two females sharing the territory.

Spotty, as I named her, appeared to be a helper in the territory; vigorously defending it with the pair. In only one year did I suspect that Spotty had her own nest (which failed). She stayed on after Rev and Speckles disappeared in early 2014 and soon attracted a mate. They fledged two chicks.

(Rev top, Speckles centre and Spotty below)

November 2014

Robin Prytherch Nov. 14

ABBY

This female buzzard has already appeared on my 2005 and 2007 cards. The 2005 card detailed the history of the Caswell Cross territory where Abby eventually settled. The 2007 card told the story of how she was caught up in a complicated scenario involving her mate Indi and two other females – a buzzard quartet.

I first saw Abby in March 2001 when she was in her second year plumage, indicating that she fledged in 1999. She defended her small territory like an adult, calling and displaying at intruding younger buzzards. As a second-year bird she looked rather paler than when she became fully adult. I show her head next to her appearance as an adult, alert and about to fly. She gained this plumage after a major moult. She was then attracted to a male in the adjacent territory, Caswell Cross.

Abby fledged her first chick in 2004. Then, in 2006, her mate, Indi, was involved with two other females, Pip and Cala (mother and daughter), in the adjacent territory, Portbury Moor. Cala disappeared at the end of 2008; the next year Abby fledged her second chick and Pip fledged two, so the male, Indi, was still holding the two territories. All change again in 2011! Both Pip and Indi disappeared when Abby paired with a new male. This year, she fledged three chicks, presumably because of the new vigorous male and no competition. Then, in her final three years, she fledged only single chicks. Confused? The lives of buzzards are full of twists and turns.

My final sighting of Abby was on 7th January 2015, and within a few days a new female appeared. Abby had lived for 15 and a half years. During her life she fledged just eight chicks – rather low productivity, but not helped by the polygamous behaviour of her first mate, Indi.

November 2015

Robin Prytherch Nov.2015

FRAN

Fran was easier to check than most buzzards because, apart from her distinctive plumage, she carried a ring on her right leg. She was probably ringed by me and my colleagues in about 1994, when we were ringing many chicks in my study area. This revealed that approximately 75% of the chicks were subsequently recovered within the study area (see Bibliography); Fran could have been one of these chicks.

I first saw Fran on 10th March 1997 sitting next to Chestnut, the resident male (shown on my 2008 card). She is a very well-marked bird and quite different from Willow, Chestnut's previous mate, who was quite pale and had died in late 1996. Up to 2009 they had fledged 13 chicks. In the next year, Chestnut went missing. He probably died when he must have been about 19 years old. He was replaced by a new male in 2010, and he and Fran fledged a further four chicks up to this year when Fran disappeared. She had fledged a total of 17 chicks in her life – an average result.

She used several nest sites, but most often the nest was in an oak or a birch. The birch nest was close to the edge of the wood in the valley bottom, and I could watch her on the nest from a nearby lane. I had my telescope mounted on my car window and could see her well until the foliage grew to block my view. She was using this nest this year and I had the feeling that she died during incubation, as a few weeks later the nest had been deserted. This year was her 20th in the territory (named Norton's Wood) so, allowing for three years 'growing-up', she must have been at least 23 years old - a long life for a buzzard.

November 2016

Robin Prytherch. Nov. 2016

CAL AND HAK

This pair was featured on my 2001 card, the year in which they arrived in the territory known as Hack's Wood (taking over the same territory previously used by Rufous and Nan). In 2014, Cal and Hak, in turn, passed the territory onto a new pair when they, presumably, died – they would have been at least 16 years old. So, Cal and Hak were together in the territory for 13 years.

For most of those years they used the same nests, in oaks, as had been used by Rufous and Nan. Cal and Hak fledged 17 chicks – they failed in just two years, and were successful in the other 11, with broods of one in six years, broods of two in four years and just one brood of three. This is an average-to-good result for a pair of buzzards in my study area in North Somerset.

Cal, illustrated adding a small stick to her nest, was featured in the latest paper of my study. It discussed the nests, nest trees and prey items brought to the nest by buzzards and was published in *British Birds* in May 2013 (see Bibliography). Nests can get quite large, but mostly they stay at about the same size, as old rotten material will fall away from the bottom of the nest. They vary in size from about 45 cm to 65 cm across the top and are approximately 28 cm deep.

November 2017

BUZZARDS HUNTING

During my studies, I am always keen to find buzzards catching or eating prey. I have recorded 35 species of prey at nests during the breeding season (including mammals, birds and amphibians). This does not include the many invertebrates that buzzards capture, especially out of the breeding season. They also steal food from other buzzards, sparrowhawks and kestrels.

The sketches on this card show a male buzzard eating a pigeon (probably a feral pigeon). I found him eating the pigeon on a fence post on 1st September 2003, at 15:05. The three sketches on the left were drawn rapidly, soon after I'd found him, and the fourth sketch was drawn just before he finished and flew off at 15:34. Whilst feeding, he sleeked his plumage, giving him a very long-legged appearance – a very typical behaviour that must help to avoid fouling of the plumage by the prey.

Another feeding incident involved a female in her territory, Clevedon Moor – just 3 km to the west – on 13th June 2015. It was 15:11 and she was on a conspicuous perch, just 100 m or so from her nest, looking very alert. She suddenly flapped off fast to the west along a small river and pounced on something (not visible to me), flushing two carrion crows, which mobbed her furiously. She was obviously 'on' something, but I could only see her back and head. Then, a grey heron appeared, walking up the riverbank to watch the action! She then, at 15:15, flapped off, with a fish in her talons, back to her nest and chicks. So, had the carrion crows stolen the fish from the grey heron, creating a scene that attracted the buzzard? It looked like that to me. I reckoned that the fish was a small eel. One rarely has the opportunity to observe such behaviour in the field. It compensates for all the hours of watching with little or no action.

October 2018

Field sketch 01/09/03, redrawn, Oct. 18
Robin Prytherch

BLONDIE

Blondie was an exceptionally pale bird. Her underparts were white with a blond tinge, hence her name. She took over the territory from another bird that presumably had died. Blondie frequently perched in a conspicuous position, such as the dead top branches of an old walnut tree. She was fully adult when I first saw her in 2011 and fledged three chicks that year.

In 2012, just one chick was fledged. Then, in 2013 and 2014, two chicks were fledged in each year. In the next two years, however, no chicks were fledged. Three chicks fledged in 2017, but the pair failed in 2018. This year, 2019, she was present early on, but then went missing. The busy A369 runs through one edge of the territory, and I wondered whether she was killed when hunting close to the road. Eleven chicks fledged from eight breeding seasons is a modest result. As far as I could tell she was mated to the same male throughout.

The territory is on the northern edge of my study area on a northern slope in a pleasant rural area, but which overlooks the industrial area of Avonmouth. Later this summer, the male was with a new female, but they did not breed.

November 2019

Robin Prytherch. 2019

BUZZARD SKETCHES

During my regular studies of buzzards in North Somerset, I occasionally get good views of birds as they hunt for prey – mostly earthworms and other invertebrates. This usually happens in autumn, especially where fields have been freshly tilled. Flight activity is not as easy to catch, since I am usually viewing the birds from within my car with the telescope mounted on the car window.

My study population has this year reached 120 pairs in 75 sq. km. This could be a ceiling, as during my regular visits to the study area I have not seen any additional pairs or single adults waiting to fill a territory. Time will tell.

November 2020

Sadly, time did not tell. This was to be Robin's last Christmas card. He died on 3rd March 2021, aged 81. These rough drawings of birds in the Flax Bourton and Westpark Wood territories date from 2004, but I have been unable to ascertain where they came from. Certainly, they don't appear in any of the notebooks or sketchbooks that came to light after he died. By this stage he was very ill, hence the short write-up and some recycled drawings from the past. However, compared with the very precise pen-and-ink illustrations that appeared on previous cards, these lightly drawn pencil sketches have a very different quality to them. Although their reproduction here is poor, they allow us to see the illustrator's impression of his subject matter in a more dynamic, lifelike way – Ed.

14.07.04 Black Kite
head & tail held up
as high as
possible
preening
15.09.04 15.35
West gate WWF go
(see note book)
29/07/04
1Y look back

BIBLIOGRAPHY

Robin Prytherch wrote countless articles in magazines and journals about birds, as well as appearing on radio and television. He also illustrated many books by other authors. Some of the published articles he wrote specifically about buzzards are listed below, together with recommended reading by other authors.

Prytherch, Robin. (1989). A survey of breeding season Sparrowhawks, Buzzards and Kestrels 1980-84. **Bristol Ornithology** 20: 167-176.

Prytherch, R.J. (2009). The social behaviour of the Common Buzzard. **British Birds** 102: 247-273.

Prytherch, Robin & Roberts, Lyndon (2012). The dispersal of Common Buzzards ringed between 1984 and 2004 in North Somerset. **Bristol Ornithology** 31: 15-27.

Prytherch, R. (2013). The breeding biology of the Common Buzzard. **British Birds** 106: 264-279.

Robin J. Prytherch & Lyndon F. Roberts (2015). Sexing of nestling Common Buzzards *Buteo buteo* in Britain. **Ringing & Migration** 30:1. 57-59.

Prytherch, R. (2016). Common Buzzard nests, nest trees and prey remains in Avon. **British Birds** 109. 256–264.

Sean Walls & Robert Kenward (2020). **The Common Buzzard.** Published by T & AD Poyser, London.

Tubbs, Colin R. (1974). **The Buzzard.** Published by David & Charles, Newton Abbot, Devon.

*A full account of RJP's other publications, illustrations and professional work can be found in **Who's Who in Ornithology** (Buckingham Press, 1997).*

*His original field notebooks and sketchpads are held by the **Bristol Regional Environmental Records Centre** (BRERC).*

BUZZARD TERRITORIES

The following list of 88 buzzard territories in RJP's study area in North Somerset was compiled from his own notes. It is not complete; he identified 120 in all. The territories are listed here in alphabetical order. The numbers in brackets are the order in which the territories were named. Also listed are the individual buzzards named by RJP in each territory (♂ = male and ♀ = female). Not all of them are mentioned in the Christmas card collection.

AWT Reserve (15) *Rev* ♂ *Speckles* ♀ *Spotty* ♀
Back Hill (67)
Backwell Common (62)
Barrow Court (56)
Barrow Court Lane (87)
Barrow Hill (54)
Baye's Wood (28) *Chinstrap* ♂
Belmont Combe (48) *Beta* ♂ *Beam* ♀
Belmont Hill (49) *Bloom* ♂ *Belle* ♀
Belmont House (*territory not numbered*) *Old Girl* ♀
Birch Wood (22)
Black Strip (2) *Rip* ♂ *Bib* ♀
Blackberry Lane (69)
Bourton Combe (Quarry) (52)
Bourton Combe (Top) (53)
Breach Wood (33)
Brook Farm (Wombles Wood) (50)
Bulling's Wood (21)
Bullock's Bottom (23)
Castle Hill (10)

Caswell Cross (73) *Cas* ♂ *Indi* ♂ *Tear* ♀ *Abby* ♀
Cherry Wood (60)
Church Wood (58)
Clapton Church (17)
Clapton Moor (82)
Clapton Wick (14) *Tertials* ♂ *Tuff* ♂ *Tess* ♀ *Tam* ♀
Conygar Quarry (11) *"Q"* ♂ *Corry* ♂ *Crescent* ♀
Copse (61)
Cosgrove Wood (57)
Court Hill (25) *Nero* ♂ *Mask* ♀
Court Wood (26) *Slot* ♂
Failand Farm (44)
Failand Lawn (77)
Failand Lodge (45)
Flax Bourton (63) *Flax* ♂ *Floss* ♀
Folly Farm (29)
Gable Wood (51) *Black Beauty* ♂ *New Boy* ♂ *Vig* ♂ *Raft* ♂ *Old Girl* ♀ *Fave* ♀
Hack's Wood (9) *Rufous* ♂ *Hak* ♂ *Buff* ♀ *Nan* ♀ *Cal* ♀
Hails Wood (78)

Happerton (39)
Horse Race (38) *Sooty* ♂ *Larch* ♀
Snowy ♀
Hyatts Wood (65)
Keedwell (66)
Kingcott Farm (79)
Leigh Court (71)
Lime Breach Wood (30)
Contrast ♂ *Comic* ♀
Lower Caswell (20) *Haze* ♂ *Ivy* ♀
Lower Failand (81)
M5 Cliffs (86)
Manor Farm (3)
Markham Brook (40)
Markham Farm (74)
Mill Combe (76)
Moat House (24)
Naish (18)
North Weston (4) *Ivory* ♂
Norton's Wood (13) *Chestnut* ♂
Willow ♀ *Fran* ♀
Old Hill (37)
Old Park (88)
Orchard (27)
Portbury Moor (72) *Hun* ♂
Pip ♀ *Cala* ♀
Research Farm (85)
Reservoir Farm (80)
Sidelands Cottages (70)
Sidelands Wood (35)

Slade Wood (55)
Stevens Farm (59)
Stone Edge Batch (84)
Three-cornered Wood (42)
Tip (64) *Tan* ♂ *Teasel* ♀
Tyntesfield East (47)
Tyntesfield West (46) ♂
Upper Caswell (19) *Curt* ♂
Como ♀ *Cora* ♀ *Tear* ♀
Vowles Bottom (41)
Walton (12) *Riv* ♂ *Honey* ♀
Walton Down (1) *Grey* ♂
Yelper ♀
Warren House Plantation (83)
West Hill (32)
Weston Big Wood (5) *Blotch* ♂
Buff2 ♀
Weston Lodge (6) *Wes* ♂ *Wyn* ♀
Weston Moor (7) *Grey Tail* ♂
Grizzle ♀
Weston Wood (8) *Dusky* ♂
Broken Wing ♀ *Dun* ♀
Westpark Wood (16)
Chocolate ♂ *Cream*
Whitehouse (31)
Windmill Hill (36)
Wraxall (34)
Wynhol (68)
Wynhol West (75)
Yew Tree Plantation (43)

INDEX

Place names mentioned in the cards are shown in **bold** and buzzard names are in *italics*. See Territories section for a more comprehensive list of locations and buzzard names.

Abby 40, 44, 60
Aerial play 24
Aggression – see 'Conflict' below
Amphibians (prey) 66
Angel posture 48
Assertive bow 48
Avonmouth 68
AWT Reserve, Gordano Valley 58
Belmont House 22
Bigamy – see Polygamy below
Birch tree 62
Birds (prey) 66
Black Beauty 20
Black-headed gull (prey/carrion) 40
Blindness 26
Blotch 50
Breeding age 38
Breeding density 70
Breeding success – see 'Fecundity' below
Buff 2 32, 50
Cainism 50
Cal 32, 64
Cala 44, 48, 60
Carrion crow 40, 66
Cas 40
Caswell Cross 40, 44, 60

Chestnut 46, 62
Chicks attacking each other 50
Chocolate 34
Clapton in Gordano 26
Clapton Wick 36
Clevedon 46
Clevedon Moor 42, 66
Como 40
Conflict 36, 38, 42, 44, 48, 50
also see 'Expelling intruders' and 'Long-staying children' below
Conygar Quarry 54
Cooperative breeding 22, 32, 44, 48, 58
Copulation 44
Cora 26, 40
Corvids (prey) 40
Court Hill 42
Cream 34
Crescent 54
Curt 26
Cuss 34
Death 20, 22, 28, 32, 34, 36, 42, 44, 52, 54, 56, 58, 60, 62, 68
Death during incubation 62
Defence of nest 58
Desertion 54, 62
Diet – see 'Prey' and 'Foraging' below

Disability 26, 46

Displacement 48

Display stoop 56

Displaying at young intruders 60

Divorce 40

Dominant posture 38, 48

Domination 40, 48

Earthworms (prey) 22, 24, 40, 42, 46, 70

Eel (prey) 66

Electrocution 46

Eviction 44, 48

Expelling intruders 24, 38, 48, 50

Fecundity 20, 26, 28, 30, 32, 34, 36, 38, 40, 44, 46, 48, 50, 52, 62, 64, 68

Fighting – see 'Conflict' above

Fish (prey) 66

Fitness 26

Flax 38

Flax Bourton 20, 38

Floss 38

Foraging 42, 46

Fratricide – see 'Cainism' above

Gatherings 38

Gordano Valley 36, 50, 52

Gos 56

Goshawk 56

Grey heron 66

Hack's Wood 22, 32, 64

Hak 32, 64

Hal 20

Haze 28

Honey 52, 56

Honeysuckle 52

Horses 46

Hun 44, 48

Hunting 66

Incest 32, 58, 60

Indi 40, 44, 48, 60

Infertility 54

Injury 26, 46

Invertebrates (prey) 42, 70

Ivy 28

Juvenile dispersal 36, 38

Kestrel 40

Larch 30

Larch (tree) 56

Living alone 40

Longevity 20, 22, 28, 32, 34, 36, 42, 46, 50, 52, 54, 60, 62, 64

Long-staying children 36, 38, 44, 46, 48, 58

Lookout perch 36, 68

Magpie 40

Mammals (prey) 66

Mask 30, 42

Mice (prey) 40

Mobbing by carrion crows 66

Mock prey 24

Moles (prey) 46

Moult 20, 60

Murder 40, 48

Nan 22, 32, 64

Nero 30, 42

Nest defence 58

Nest failure 42, 46, 50, 56, 58, 68

Nest structure 56, 64

Nest use 54, 64

New Boy 20

Norton's Wood 46, 62

Oak (tree) 62, 64

Old Girl 20, 22

Peregrine 40

Pigeons (prey) 40
Pip 44, 48, 60
Piracy 40, 56, 66
Plumage 28, 30, 42, 46, 54, 56, 60, 68
Polygamy 22, 44, 52, 56, 60
Poplar (tree) 36
Portbury 48
Portbury Moor 44, 60
Prey 22, 24, 40, 42, 46, 66, 70
Q 54
Rabbits (prey) 40
Riv 52, 56
Rook 26
Rufous 22, 30, 32, 64
Sibling rivalry in nest 50
Signalling 48
Sleeked plumage 66
Snowy 30
Sooty 30
Sparrowhawk 40
Stealing food – see 'Piracy' above

Study population 66
Submissive behaviour 48
Tear 40
Territorial defence 44, 48, 50, 62
Territory takeover 44
Territory size 52, 54
Tertials 36
Tess 36
Theft of prey – see 'Piracy' above
Thrushes (prey) 40
Tits (prey) 40
Training for adult life 24
Trio of adults in territory 46
Tuff 36
Two females tending one nest 44
Vig 20
Voles (prey) 40
Weston Big Wood 50
Westpark Wood 34
Willow 46, 62
Willow (tree) 26

ACKNOWLEDGEMENTS

I would like to thank Robin Prytherch's near family and his executors for allowing me to publish this book.

I would like to thank Joe Burt and Clive Burlton at Bristol Books for agreeing to take on this project. Their input at every stage has been invaluable.

I also wish to thank all the sponsors whose generous contributions made this publication possible.

Thanks are due to those who kindly gave their time and applied their thought to arriving at a suitable format and content for this book, for commenting on pre-publication drafts and providing material, in particular – Judy Copeland, Dan Freeman, Ken Hall, Hilary Jeffkins, Rosamund Kidman Cox, Mike Lord and Di Williams.

Copywriters Lisa Benjamin and Mark Binnersley made many useful suggestions that improved the introductory sections of the book; Hazel Atashroo provided a drawing of Robin that appears on the dust cover; Andrew Jakubowski helped to improve the digital reproduction of the final drawing (Buzzard Sketches); and Miranda Krestovnikoff assisted with the launch and promotion of this publication. I am grateful to them all.

Finally, I would like to thank Robin himself for introducing me to the fascinating world of buzzards and teaching me how to observe them.

A NOTE ABOUT OUR SPONSORS

The Bristol Naturalists' Society

The Bristol Naturalists' Society exists to stimulate a greater awareness of natural history and geology in the Bristol area. The Society was founded in 1862. Robin Prytherch served on the Committee of the Ornithology Section in the 1960s. The Society is actively involved in research and conservation. Each year its talks, trips and publications are enjoyed by hundreds of people wanting to find out more about our natural world.

Bristol Ornithological Club

The Bristol Ornithological Club (BOC) was founded, in 1966, to promote, encourage and co-ordinate the scientific study of ornithology in all its branches in the Bristol area. Robin Prytherch was a founder member of the Club and served as its Chairman in the 1970s, as well as being a Committee member for nearly 50 years. He also designed the Club's logo.

Contributions were also received from:

Michael Allen; Mark Avery; Mike Bailey; Mike Barton; Peter Belcher; Dick Best; Mark Binnersley; *British Birds*; Sue Black; Ian Carter; Richard Chandler; *Chew Valley Ringing Station*; David & Shirley Clegg; Theo Cockerell & Monika Shove; Rob Collis; Judy Copeland; David Davies; Tim Dee; Phyl Dykes; Colin Everett; Alan Feest; Dan & Becky Freeman; Helen Gilks; Aurora Gonzalo-Tarodo; Margaret Gorely; Ken & Lys Hall; Jessica Holm; April & Lawrence Humphreys; John Hutchinson; Pamela Jackson; Gareth Jones; Rosamund Kidman Cox; Roderick Leslie; Mike & Alix Lord; Vanessa Manceron; Bob Medland; Nigel Milbourne; Sally & Ron Moore; Stephen Moss; Ian Packer; Vassili Papastavrou; Tony Parsons; Brian Rabbitts; Roger Riddington; Lyndon & Sally Roberts; Harvey Rose; Tony & Hilary Soper; Chris Sperring MBE; Eve Tigwell; Andrew Waygood; Robert & Catherine Webber; and Di Williams.